Essential Concepts in Bacteriology

A Primer for Healthcare Professionals

Bhupen Thapa

Copyright © 2024 Bhupen Thapa
All rights reserved.
No part of this book may be reproduced or transmitted in any form or by any means, electronic or mechanical, including photocopying, recording, or by any information storage and retrieval system, without permission in writing from the copyright holder.

Table of Contents

Chapter 1: Introduction to Bacteriology 1
   The History of Bacteriology 1
   The Scope of Bacteriology 2
   The Importance of Bacteriology in Healthcare 3
Chapter 2: Bacterial Structure and Function 4
   Cell Envelope 4
   Cell Wall 5
   Cell Membrane 6
   Cytoplasm 6
   Ribosomes 7
   Flagella and Pili 8
   Capsules and Slime Layers 9
Chapter 3: Bacterial Genetics 11
   DNA Replication 11
   Transcription 12
   Translation 13
   Mutation and Genetic Variation 14
   Horizontal Gene Transfer 14
Chapter 4: Bacterial Growth and Metabolism 16
   Nutritional Requirements 16
   Bacterial Growth Curve 17
   Metabolic Pathways 18
   Energy Production 19
Chapter 5: Bacterial Pathogenesis 20
   Host-Pathogen Interactions 20
   Virulence Factors 21
   Mechanisms of Pathogenicity 21
   Immune Responses to Bacterial Infections 22

Chapter 6: Principles of Bacterial Identification and Classification 24
    Morphological and Biochemical Characteristics 24
    Molecular Techniques 25
    Antimicrobial Susceptibility Testing 26
Chapter 7: Antibiotics and Antibiotic Resistance 28
    Mechanisms of Action 28
    Antibiotic Classes 29
    Factors Contributing to Antibiotic Resistance 30
    Strategies to Combat Antibiotic Resistance 31
Chapter 8: Bacterial Diseases 32
    Gram-Positive Bacterial Infections 32
    Gram-Negative Bacterial Infections 33
    Atypical Bacterial Infections 34
    Emerging Infectious Diseases 35
Chapter 9: Bacterial Control and Prevention 36
    Sterilization and Disinfection 36
    Hand Hygiene 37
    Vaccination 38
    Infection Control Measures 38
Chapter 10: Future Directions in Bacteriology 40
    Advances in Bacterial Research 40
    Potential Therapeutic Strategies 41
    Global Health Implications of Bacterial Infections 42
Chapter 11: Conclusion 43
    Summary of Key Concepts 43
    Recommendations for Further Reading 44
    Importance of Bacteriology in Healthcare Practice 45
Dear Reader, ... 47

A Primer for Healthcare Professionals

Chapter 1
Introduction to Bacteriology

The History of Bacteriology

Bacteriology is a field of science that focuses on the study of bacteria, microscopic organisms that play a significant role in both the environment and human health. The history of bacteriology dates back to the late 17th century when Anton van Leeuwenhoek first observed bacteria under a microscope. His groundbreaking discovery paved the way for further research into these tiny organisms and their impact on the world around us.
In the 19th century, Louis Pasteur and Robert Koch made significant contributions to the field of bacteriology. Pasteur is known for his work on the germ theory of disease, which revolutionized our understanding of how bacteria cause infections. Koch, on the other hand, developed techniques for isolating and identifying disease-causing bacteria, laying the foundation for the field of medical microbiology.
The discovery of antibiotics in the early 20th century marked a major turning point in the history of bacteriology. Alexander Fleming's discovery of penicillin in 1928 revolutionized the treatment of bacterial infections and saved countless lives. The development of other antibiotics, such as streptomycin and tetracycline, further expanded our ability to combat bacterial diseases.
As our knowledge of bacteria has grown, so too has our understanding of their role in both health and disease. Bacteriology has played a crucial role in the development of vaccines, the study of antibiotic resistance, and the investigation of infectious diseases such as tuberculosis and cholera. Today, bacteriology continues to be a vibrant and dynamic field of study, with new discoveries and advancements being made every day.
In conclusion, the history of bacteriology is a testament to the power of scientific inquiry and discovery. From the early observations of Leeuwenhoek to the groundbreaking work of Pasteur and Koch, the field has come a long way in understanding the role of bacteria in our world. As we continue to unravel the mysteries of these tiny organisms, we are better equipped to prevent and treat bacterial infections and improve human health.

The Scope of Bacteriology

Bacteriology is a branch of microbiology that focuses on the study of bacteria, their structure, physiology, genetics, and their role in causing diseases. The scope of bacteriology is vast and encompasses a wide range of topics that are essential for understanding the role of bacteria in healthcare. In this subchapter, we will explore the scope of bacteriology and its significance in the field of healthcare.

One of the key aspects of bacteriology is the study of bacterial morphology and structure. By understanding the physical characteristics of bacteria, scientists can classify and identify different species of bacteria. This knowledge is crucial for diagnosing bacterial infections and determining the appropriate treatment. Additionally, studying bacterial structure can provide insights into how bacteria interact with their environment and host organisms.

Another important aspect of bacteriology is the study of bacterial metabolism and physiology. Bacteria are incredibly diverse in their metabolic capabilities, with some species being able to thrive in extreme environments while others are pathogenic and cause disease in humans. Understanding bacterial metabolism is essential for developing new antibiotics and treatments for bacterial infections.

Bacteriology also plays a crucial role in the field of epidemiology and public health. By studying how bacteria spread and cause disease, scientists can develop strategies to prevent the spread of infections and control outbreaks. This includes monitoring antibiotic resistance in bacteria and developing vaccines to prevent bacterial infections.

In addition to its role in healthcare, bacteriology also has applications in biotechnology and environmental science. Bacteria are used in a variety of biotechnological processes, such as the production of antibiotics, enzymes, and other useful products. Bacteriology also plays a key role in studying the role of bacteria in environmental processes, such as nutrient cycling and bioremediation.

Overall, the scope of bacteriology is vast and encompasses a wide range of topics that are essential for understanding the role of bacteria in healthcare and other fields. By studying bacteria at a molecular level, scientists can gain valuable insights into how these microorganisms function and interact with their environment, ultimately leading to advancements in healthcare, biotechnology, and environmental science.

The Importance of Bacteriology in Healthcare

Bacteriology is a crucial field of study in healthcare, as it deals with the identification and understanding of bacteria, which are microscopic organisms that can cause a wide range of diseases in humans. By studying bacteriology, healthcare professionals can better diagnose and treat bacterial infections, leading to improved patient outcomes and overall public health. In this subchapter, we will explore the importance of bacteriology in healthcare and how it affects medical practice.

One of the key reasons why bacteriology is so important in healthcare is its role in the prevention and control of infectious diseases. By understanding the characteristics of different bacteria and how they spread, healthcare professionals can develop strategies to prevent outbreaks and limit the spread of infections. This knowledge is crucial in settings such as hospitals and long-term care facilities, where vulnerable populations are at a higher risk of contracting infections.

Bacteriology also plays a crucial role in the field of antibiotic resistance. As bacteria evolve and develop resistance to antibiotics, it is essential for healthcare professionals to stay informed about the latest developments in bacteriology to ensure that effective treatments are available for patients. By understanding how bacteria develop resistance and how to combat it, healthcare professionals can help prevent the spread of drug-resistant infections and improve patient outcomes.

Furthermore, bacteriology is essential for the development of new treatments and vaccines for bacterial infections. By studying the biology and behavior of bacteria, researchers can identify potential targets for drug development and design effective vaccines to prevent infections. This research is crucial in the fight against emerging infectious diseases and can lead to significant advancements in medical practice.

Overall, the importance of bacteriology in healthcare cannot be overstated. By studying bacteria and their impact on human health, healthcare professionals can better understand, diagnose, and treat bacterial infections. This knowledge is crucial for preventing outbreaks, controlling antibiotic resistance, and developing new treatments and vaccines. As our understanding of bacteriology continues to evolve, so too will our ability to combat infectious diseases and improve patient care.

Chapter 2
Bacterial Structure and Function

Cell Envelope

The cell envelope is a crucial structure in bacteria that serves as a protective barrier against external threats and helps maintain the integrity of the cell. Composed of several layers, the cell envelope consists of the cell membrane, cell wall, and outer membrane in Gram-negative bacteria. Understanding the composition and function of the cell envelope is essential for healthcare professionals, as it plays a significant role in bacterial pathogenesis and antibiotic resistance.

The cell membrane, also known as the plasma membrane, is the innermost layer of the cell envelope and is responsible for controlling the movement of molecules in and out of the cell. It is composed of a phospholipid bilayer embedded with proteins that perform various functions, such as nutrient transport and energy production. The cell membrane is also involved in cell signaling and communication with other cells, making it a critical component of bacterial survival and growth.

The cell wall is the middle layer of the cell envelope and provides structural support and shape to the bacterial cell. In Gram-positive bacteria, the cell wall is composed of a thick layer of peptidoglycan, a unique polymer made up of amino acids and sugars. This layer helps protect the cell from osmotic pressure and provides resistance to mechanical stress. In contrast, Gram-negative bacteria have a thinner layer of peptidoglycan sandwiched between an inner cell membrane and an outer membrane.

The outer membrane is a unique feature of Gram-negative bacteria that provides an additional barrier against external threats, such as antibiotics and immune system defenses. Composed of lipopolysaccharides (LPS) and proteins, the outer membrane acts as a selective barrier that allows only certain molecules to pass through. This outer layer also plays a role in bacterial virulence, as it contains components that can interact with host cells and modulate the immune response.

Overall, the cell envelope is a complex and dynamic structure that is essential for bacterial survival and pathogenesis. By understanding the composition and function of the cell envelope, healthcare professionals can better combat bacterial infections and develop more effective treatment strategies. Additionally, research on the cell envelope is ongoing, as scientists continue to uncover new insights into bacterial physiology and antibiotic resistance mechanisms.

Cell Wall

The cell wall is a crucial component of bacterial cells that plays a vital role in maintaining the structural integrity and shape of the cell. It is a rigid, protective layer that surrounds the bacterial cell membrane, providing support and protection against environmental stresses. The composition of the cell wall varies among different bacterial species, but it is generally composed of peptidoglycan, a unique molecule that is not found in eukaryotic cells.

Peptidoglycan is a complex polymer made up of repeating units of sugars and amino acids that are cross-linked together to form a strong mesh-like structure. This structure gives the cell wall its rigidity and strength, allowing it to withstand changes in osmotic pressure and protect the cell from bursting. The thickness and composition of the peptidoglycan layer can vary between bacterial species, with some bacteria having thicker cell walls than others.

In addition to providing structural support, the cell wall also plays a crucial role in the interaction between bacteria and their environment. The cell wall contains various proteins and other molecules that are involved in processes such as nutrient uptake, cell signaling, and adhesion to surfaces. These interactions are essential for the survival and growth of bacteria in their natural habitats.

One of the key characteristics of the cell wall is its ability to be targeted by antibiotics. Many antibiotics work by disrupting the synthesis or integrity of the bacterial cell wall, leading to cell death. This makes the cell wall an attractive target for antimicrobial therapy and has led to the development of a wide range of antibiotics that specifically target the cell wall of bacteria.

Understanding the structure and function of the cell wall is essential for healthcare professionals involved in the treatment of bacterial infections. By targeting the cell wall, antibiotics can effectively kill bacteria and treat infections, highlighting the importance of this fundamental component of bacterial cells. As research continues to uncover new insights into the cell wall and its role in bacterial physiology, healthcare professionals can continue to develop more effective strategies for combating bacterial infections.

Cell Membrane

The cell membrane is a critical component of bacterial cells, serving as a barrier that separates the interior of the cell from the external environment. Made up of a phospholipid bilayer embedded with proteins, the cell membrane is selectively permeable, allowing only certain substances to pass through while blocking others. This selective permeability is essential for maintaining the integrity of the cell and regulating the flow of nutrients and waste products.

One of the key functions of the cell membrane is to regulate what enters and exits the cell. This is achieved through the presence of transport proteins that span the membrane and facilitate the movement of molecules across the lipid bilayer. Some transport proteins act as channels, allowing specific substances to pass through, while others function as carriers, binding to specific molecules and transporting them across the membrane.

In addition to regulating the movement of molecules, the cell membrane also plays a role in cell signaling and communication. Receptor proteins located on the surface of the membrane bind to specific signaling molecules, triggering a series of events within the cell that can lead to changes in gene expression or cellular behavior. This signaling process is crucial for coordinating the activities of individual cells and enabling them to respond to changes in their environment.

The composition of the cell membrane can vary between different types of bacteria, with some species possessing additional layers such as a cell wall or capsule. These additional layers can provide extra protection against environmental stresses or help the bacteria adhere to surfaces. Understanding the structure and function of the cell membrane is essential for studying bacterial physiology and developing new strategies for treating bacterial infections.

Overall, the cell membrane is a dynamic and complex structure that plays a vital role in the survival and function of bacterial cells. By studying the composition and function of the cell membrane, researchers and healthcare professionals can gain valuable insights into the mechanisms of bacterial pathogenesis and identify potential targets for antimicrobial therapies.淵

Cytoplasm

Cytoplasm is a crucial component of bacterial cells, serving as the site for many essential cellular processes. It is a gel-like substance that fills the interior of the cell, surrounding the genetic material and organelles.

The cytoplasm is composed of water, proteins, nucleic acids, ions, and other molecules necessary for the cell's survival and function.

One of the primary functions of cytoplasm is to house the cell's genetic material, including the DNA and RNA necessary for protein synthesis. Within the cytoplasm, ribosomes play a key role in translating the genetic code into functional proteins. Additionally, enzymes and other proteins within the cytoplasm carry out metabolic processes such as energy production and nutrient metabolism.

Cytoplasm also serves as a storage site for various molecules, including nutrients and waste products. This allows the cell to regulate its internal environment and respond to changing external conditions. In addition, the cytoplasm provides structural support for the cell, helping to maintain its shape and integrity.

The cytoplasm is a dynamic environment that is constantly changing in response to the cell's needs. It is involved in processes such as cell division, movement, and communication with other cells. As such, understanding the role of cytoplasm in bacterial cells is essential for healthcare professionals working in the field of bacteriology.

In conclusion, cytoplasm is a vital component of bacterial cells that plays a central role in many cellular processes. Its composition and function are critical for the survival and function of the cell. By studying the properties of cytoplasm and how it interacts with other cellular components, healthcare professionals can gain valuable insights into bacterial physiology and pathogenesis.

Ribosomes

Ribosomes are essential components of bacterial cells that play a crucial role in protein synthesis. These small, complex structures are responsible for translating the genetic information stored in the bacterial DNA into functional proteins. Ribosomes are found in all living cells, including bacteria, and are often referred to as the "protein factories" of the cell.

In bacterial cells, ribosomes are composed of two subunits - the large subunit and the small subunit. These subunits work together to read the messenger RNA (mRNA) that carries the genetic code for a specific protein. The ribosomes then assemble amino acids in the correct order to form the protein according to the instructions encoded in the mRNA.

The process of protein synthesis begins when the ribosome binds to an mRNA molecule at a specific sequence called the start codon. This signals the ribosome to begin reading the genetic code and assembling the corresponding amino acids. As the ribosome moves along the mRNA molecule, it reads each codon and adds the corresponding amino acid to the growing protein chain.

Ribosomes are essential for the survival of bacterial cells, as proteins are necessary for almost every cellular function. Without ribosomes, bacteria would not be able to produce the enzymes, structural proteins, and other molecules needed for growth and metabolism. Understanding the structure and function of ribosomes is crucial for healthcare professionals studying bacteriology, as it provides insight into how bacteria survive and thrive in different environments.

In conclusion, ribosomes are integral components of bacterial cells that are responsible for protein synthesis. These complex structures play a vital role in translating the genetic information stored in bacterial DNA into functional proteins. Healthcare professionals studying bacteriology must have a solid understanding of ribosomes and their function to better comprehend how bacteria grow, replicate, and cause disease.

Flagella and Pili

Flagella and pili are two important structures found on the surface of bacteria that play crucial roles in their movement and adhesion. Flagella are long, whip-like appendages that allow bacteria to move in a fluid environment. They are made up of a protein called flagellin and are attached to the cell membrane through a complex motor system. The rotation of flagella propels the bacterium forward, allowing it to swim towards nutrients or away from harmful substances.

Pili, on the other hand, are short, hair-like structures that extend from the surface of the bacterium. They are made up of a protein called pilin and are involved in attachment to surfaces or other bacteria. Pili help bacteria adhere to host cells and form biofilms, which are communities of bacteria that stick together and are often resistant to antibiotics. Pili also play a role in the transfer of genetic material between bacteria through a process called conjugation.

Both flagella and pili are important virulence factors that contribute to the pathogenicity of bacteria. For example, flagella allow bacteria to move towards and invade host cells, leading to infection. Pili help bacteria adhere to host tissues and evade the immune system, increasing their ability to cause disease. Understanding the structure and function of flagella and pili is essential for developing strategies to combat bacterial infections and prevent the spread of antibiotic-resistant strains.

In addition to their role in pathogenicity, flagella and pili are also important in the identification and classification of bacteria. Different species of bacteria have unique flagellar and pili arrangements, which can be used to distinguish between them. This information is valuable in clinical settings for diagnosing infections and choosing appropriate treatments. By studying flagella and pili, healthcare professionals can gain insights into the behavior and characteristics of bacteria, ultimately improving patient care and public health outcomes.

In conclusion, flagella and pili are fascinating structures that play key roles in the movement, adhesion, and pathogenicity of bacteria. Understanding the structure and function of these appendages is essential for healthcare professionals working in the field of bacteriology. By studying flagella and pili, we can gain valuable insights into the behavior of bacteria, leading to improved diagnostic and treatment strategies for bacterial infections.

Capsules and Slime Layers

In the world of bacteriology, capsules and slime layers play a crucial role in the survival and virulence of many bacterial species. These structures are composed of complex polysaccharides that surround the bacterial cell and provide protection from environmental stresses, such as desiccation and immune system attacks. Understanding the function and importance of capsules and slime layers is essential for healthcare professionals in order to better combat bacterial infections.

Capsules are tightly bound, well-organized structures that are often visible under a microscope as a distinct halo surrounding the bacterial cell. These structures are important for protecting the bacterium from phagocytosis by immune cells, as well as from desiccation and other environmental stresses. Capsules also play a role in the adherence of bacteria to surfaces, which can contribute to the formation of biofilms and the colonization of host tissues.

Slime layers, on the other hand, are loosely bound structures that are often less well-defined than capsules. Slime layers are composed of exopolysaccharides that are secreted by the bacterial cell and form a viscous layer around the cell. This layer can help bacteria adhere to surfaces and form biofilms, which are communities of bacteria embedded in a matrix of extracellular polymeric substances. Biofilms are notoriously difficult to eradicate and are a common cause of chronic infections.

Both capsules and slime layers are important virulence factors for many pathogenic bacteria. These structures can help bacteria evade the immune system, resist antibiotic treatment, and survive in hostile environments. Understanding the role of capsules and slime layers in bacterial pathogenesis is essential for developing effective strategies to combat bacterial infections and prevent the spread of antibiotic-resistant strains.

In conclusion, capsules and slime layers are important structures that play a critical role in the survival and virulence of many bacterial species.

A Primer for Healthcare Professionals

Healthcare professionals must be aware of the function and importance of these structures in order to effectively treat bacterial infections and prevent the spread of antibiotic resistance. By targeting capsules and slime layers as potential therapeutic targets, researchers may be able to develop new and innovative strategies for combating bacterial pathogens.

Chapter 3
Bacterial Genetics

DNA Replication

DNA replication is a fundamental process that occurs in all living organisms, including bacteria. This process is essential for the transmission of genetic information from one generation to the next. In bacteria, DNA replication is a highly regulated and complex process that involves the coordinated action of numerous enzymes and proteins.

The first step in DNA replication is the unwinding of the double-stranded DNA molecule. This is accomplished by an enzyme called helicase, which breaks the hydrogen bonds between the two strands of DNA, allowing them to separate. Once the DNA has been unwound, another enzyme called DNA polymerase can begin the process of copying each strand of DNA.

DNA polymerase is the enzyme responsible for synthesizing new DNA strands during replication. It does this by adding nucleotide building blocks to the growing DNA strand in a specific order dictated by the template DNA strand. DNA polymerase can only add nucleotides in the 5' to 3' direction, meaning that DNA replication occurs in a continuous manner on one strand (the leading strand) and in a discontinuous manner on the other strand (the lagging strand).

The process of DNA replication is a highly accurate and efficient process, with an error rate of less than one mistake per billion nucleotides added. This high level of fidelity is achieved through the action of DNA polymerase, which has a proofreading function that can correct errors as they occur. In addition, the DNA replication process is highly regulated, with checkpoints that ensure that each step is completed correctly before the next step can proceed.

Overall, DNA replication is a crucial process in bacteria that ensures the faithful transmission of genetic information from one generation to the next. Understanding the intricacies of this process is essential for healthcare professionals working in the field of bacteriology, as it provides insights into how bacteria evolve and adapt to changing environments. By studying DNA replication, researchers can gain a deeper understanding of bacterial genetics and potentially develop new strategies for combating bacterial infections.

Transcription

Transcription is a fundamental process in the field of bacteriology that involves the synthesis of RNA molecules from a DNA template. This process is essential for the expression of genes and the production of proteins that are vital for the survival and function of bacterial cells. Understanding the mechanisms and regulation of transcription is crucial for healthcare professionals who work with bacteria, as it provides insights into how bacterial cells function and respond to their environment.

In bacteria, transcription is carried out by RNA polymerase, an enzyme that catalyzes the formation of RNA molecules using a DNA template. The process of transcription involves several steps, including initiation, elongation, and termination. During initiation, RNA polymerase binds to a specific region of DNA called the promoter, which signals the start of transcription. Elongation involves the synthesis of an RNA molecule complementary to the DNA template, while termination marks the end of transcription and the release of the RNA molecule.

The regulation of transcription in bacteria is a complex process that involves the interaction of various regulatory elements, including transcription factors, promoters, and repressors. These elements play a crucial role in controlling when and where transcription occurs, allowing bacterial cells to respond to changes in their environment and regulate the expression of specific genes. Understanding the regulatory mechanisms of transcription is important for healthcare professionals, as it can provide insights into how bacteria adapt to different conditions and develop resistance to antibiotics.

Transcription is a highly dynamic process that can be influenced by a variety of factors, including environmental conditions, cellular signaling pathways, and the presence of specific molecules. Changes in transcriptional activity can have profound effects on the function and behavior of bacterial cells, impacting their ability to survive, reproduce, and cause disease. Healthcare professionals who study bacteriology must have a thorough understanding of transcription and its regulation in order to effectively diagnose and treat bacterial infections.

In conclusion, transcription is a critical process in bacteriology that plays a key role in gene expression and protein synthesis in bacterial cells. Understanding the mechanisms and regulation of transcription is essential for healthcare professionals who work with bacteria, as it provides insights into how bacterial cells function and respond to their environment. By studying transcription, healthcare professionals can gain a deeper understanding of bacterial biology and develop more effective strategies for combating bacterial infections.

Translation

Translation is a crucial process in bacteriology that involves the conversion of genetic information from DNA into proteins. This process is essential for the survival and function of bacteria, as proteins are the building blocks of cells and carry out various functions within the organism. Understanding how translation works is key to comprehending how bacteria function and how they can be targeted for treatment.

In bacteria, translation occurs in the ribosome, a complex molecular machine that reads the genetic code stored in messenger RNA (mRNA) and uses it to assemble proteins. The process begins with the initiation of translation, where the ribosome binds to the start codon on the mRNA and forms a functional complex with transfer RNA (tRNA) molecules carrying amino acids. These tRNA molecules recognize specific codons on the mRNA and bring the corresponding amino acids to the ribosome, where they are joined together to form a protein chain.

During the elongation phase of translation, the ribosome moves along the mRNA, reading each codon and adding the appropriate amino acid to the growing protein chain. This process continues until a stop codon is reached, signaling the termination of translation and the release of the completed protein. The accuracy and efficiency of translation are crucial for the proper functioning of bacteria, as errors in protein synthesis can lead to the production of non-functional or harmful proteins.

Several factors can influence the process of translation in bacteria, including the availability of amino acids, the concentration of ribosomes, and the presence of regulatory molecules that can modulate gene expression. Understanding these factors can provide insights into how bacteria respond to changes in their environment and how they adapt to different conditions. By studying the mechanisms of translation in bacteria, researchers can identify potential targets for new antibiotics and develop strategies to combat antibiotic resistance.

In conclusion, translation is a fundamental process in bacteriology that plays a critical role in the survival and function of bacteria. By understanding how translation works and the factors that influence it, healthcare professionals can gain valuable insights into the biology of bacteria and the mechanisms of antibiotic resistance. This knowledge can inform the development of new treatment strategies and help combat the growing threat of drug-resistant bacterial infections.

Mutation and Genetic Variation

Mutation and genetic variation are crucial concepts in bacteriology that play a significant role in the evolution and adaptation of bacteria. Mutations are changes in the genetic material of an organism that can result in new traits or characteristics. These changes can occur spontaneously or be induced by external factors such as radiation or chemicals. Genetic variation refers to the diversity of genetic material within a population, which can lead to differences in traits and behaviors among individuals.

Mutations can have both positive and negative effects on bacteria. Positive mutations can confer advantages such as antibiotic resistance, allowing bacteria to survive in hostile environments. On the other hand, negative mutations can result in decreased fitness or viability. Understanding how mutations occur and their impact on bacterial populations is essential for healthcare professionals to effectively combat antibiotic resistance and other bacterial infections.

Genetic variation within bacterial populations is influenced by several factors, including horizontal gene transfer, genetic recombination, and natural selection. Horizontal gene transfer involves the exchange of genetic material between different bacteria, leading to the spread of antibiotic resistance genes and other traits. Genetic recombination is the process by which genetic material is exchanged between two different organisms, resulting in new combinations of genes. Natural selection acts on genetic variation by favoring individuals with traits that provide a survival advantage in a given environment.

The study of mutation and genetic variation in bacteria is critical for understanding the mechanisms of bacterial evolution and adaptation. By investigating how mutations arise and spread within bacterial populations, researchers can develop strategies to combat antibiotic resistance and other public health threats. Healthcare professionals must be aware of the importance of genetic variation in bacteria and its implications for patient care and treatment. By staying informed about the latest research in bacteriology, healthcare professionals can make informed decisions to improve patient outcomes and public health.

Horizontal Gene Transfer

Horizontal gene transfer is a process by which bacteria can exchange genetic material with other bacteria of the same or different species. This phenomenon plays a crucial role in bacterial evolution and adaptation to new environments. Unlike vertical gene transfer, which occurs through the transfer of genetic material from parent to offspring, horizontal gene transfer allows bacteria to acquire new genes from other bacteria in their environment.

There are three main mechanisms of horizontal gene transfer in bacteria: transformation, transduction, and conjugation. Transformation involves the uptake of free DNA from the environment by a bacterial cell. This DNA can then be incorporated into the bacterial genome through recombination. Transduction occurs when a bacteriophage (a virus that infects bacteria) transfers bacterial DNA from one cell to another. Conjugation is a process in which two bacteria physically connect and transfer genetic material through a tube-like structure called a pilus.

Horizontal gene transfer can have significant implications for bacterial pathogenicity and antibiotic resistance. For example, the transfer of genes encoding virulence factors can allow non-pathogenic bacteria to become pathogenic. Additionally, the transfer of antibiotic resistance genes can lead to the spread of multidrug-resistant bacteria, making infections more difficult to treat. Understanding the mechanisms of horizontal gene transfer is essential for developing strategies to combat the spread of antibiotic resistance.

Research into horizontal gene transfer has also revealed its potential applications in biotechnology and genetic engineering. By harnessing the natural ability of bacteria to exchange genetic material, scientists can introduce new genes into bacteria for the production of valuable compounds or the degradation of environmental pollutants. Horizontal gene transfer has also been used in the development of genetically modified organisms (GMOs) for agricultural purposes.

In conclusion, horizontal gene transfer is a fascinating process that allows bacteria to rapidly adapt to changing environments through the acquisition of new genetic material. While this phenomenon can have negative consequences in terms of antibiotic resistance and pathogenicity, it also offers exciting possibilities for biotechnology and genetic engineering. By studying horizontal gene transfer, researchers can gain valuable insights into bacterial evolution and develop innovative solutions to combat antibiotic resistance and other challenges in bacteriology.

Chapter 4
Bacterial Growth and Metabolism

Nutritional Requirements

Nutritional requirements play a crucial role in the growth and survival of bacteria. Bacteria are diverse organisms that require a variety of nutrients to thrive. Understanding the nutritional requirements of bacteria is essential for healthcare professionals as it can help in the prevention and treatment of bacterial infections. In this subchapter, we will explore the key nutritional requirements of bacteria and how they impact bacterial growth and survival.

One of the most important nutritional requirements for bacteria is a source of carbon. Bacteria are classified based on their carbon source as either autotrophs or heterotrophs. Autotrophic bacteria can synthesize their own organic compounds from inorganic substances like carbon dioxide, while heterotrophic bacteria rely on organic compounds as their carbon source. The type of carbon source a bacterium utilizes can impact its growth rate and metabolism.

Another essential nutritional requirement for bacteria is a source of energy. Bacteria obtain energy through various metabolic processes such as fermentation, respiration, and photosynthesis. The type of energy source a bacterium uses can vary depending on its metabolic capabilities. For example, some bacteria can use sunlight as an energy source through photosynthesis, while others rely on organic compounds like glucose for energy production.

In addition to carbon and energy sources, bacteria also require essential nutrients like nitrogen, phosphorus, sulfur, and trace elements for growth and metabolism. These nutrients are important for the synthesis of essential biomolecules like proteins, nucleic acids, and cell membranes. Bacteria have evolved various mechanisms to acquire these nutrients from their environment, including the utilization of specialized transport systems and metabolic pathways.

Understanding the nutritional requirements of bacteria is crucial for healthcare professionals as it can help in the development of effective treatment strategies for bacterial infections. By targeting the specific nutritional needs of pathogenic bacteria, healthcare professionals can design antimicrobial therapies that inhibit bacterial growth and survival. Additionally, knowledge of bacterial nutritional requirements can aid in the development of novel antimicrobial agents that target essential metabolic pathways in bacteria.

In conclusion, nutritional requirements are essential for the growth and survival of bacteria. By understanding the diverse nutritional needs of bacteria, healthcare professionals can better combat bacterial infections and develop effective treatment strategies. Further research into bacterial nutritional requirements is crucial for the development of novel antimicrobial therapies and the prevention of antibiotic resistance.

Bacterial Growth Curve

Bacterial growth curve is a fundamental concept in bacteriology that describes the typical pattern of bacterial population growth over time. Understanding the bacterial growth curve is essential for healthcare professionals as it provides insight into the dynamics of bacterial populations and aids in the development of effective treatment strategies. The bacterial growth curve consists of four distinct phases: lag phase, exponential (log) phase, stationary phase, and death phase. During the lag phase, bacteria adapt to their new environment and prepare for rapid growth. This phase is characterized by a period of slow or no growth as bacteria synthesize enzymes and other molecules necessary for replication.

The exponential phase is when bacteria experience rapid growth and reproduction. During this phase, the population of bacteria doubles at regular intervals, leading to a logarithmic increase in cell numbers. This phase is crucial for studying the growth kinetics of bacteria and determining factors that influence bacterial growth rates.

The stationary phase occurs when the growth rate of bacteria slows down and the number of live bacteria remains relatively constant. This phase is reached when the rate of cell division equals the rate of cell death, resulting in a plateau in bacterial population growth. Factors such as limited nutrients, accumulation of waste products, and environmental stressors contribute to the onset of the stationary phase.

The death phase is the final stage of the bacterial growth curve, where the number of live bacteria decreases due to cell death exceeding cell division. This phase is inevitable for all bacterial populations and is influenced by factors such as antimicrobial treatment, environmental conditions, and genetic mutations. Understanding the bacterial growth curve is essential for healthcare professionals to develop effective treatment strategies and control measures to combat bacterial infections.

Metabolic Pathways

Metabolic pathways are essential processes that occur within bacteria to ensure their survival and growth. These pathways involve a series of chemical reactions that convert nutrients into energy, building blocks, and waste products. Understanding these metabolic pathways is crucial for healthcare professionals to better comprehend how bacteria function and how they can be targeted for treatment.

One of the most well-known metabolic pathways in bacteria is glycolysis. This pathway involves the breakdown of glucose to produce energy in the form of ATP. Glycolysis occurs in the cytoplasm of bacterial cells and is a universal pathway that is conserved across different bacterial species. By studying glycolysis, healthcare professionals can gain insight into how bacteria obtain energy to support their growth and reproduction.

Another important metabolic pathway in bacteria is the citric acid cycle, also known as the Krebs cycle. This pathway takes place in the mitochondria of eukaryotic cells but occurs in the cytoplasm of bacterial cells. The citric acid cycle is responsible for further breaking down glucose and other nutrients to produce more ATP and other molecules that are essential for bacterial survival. Understanding the citric acid cycle can provide healthcare professionals with valuable information on how bacteria regulate their metabolism to adapt to different environmental conditions.

In addition to glycolysis and the citric acid cycle, bacteria also utilize other metabolic pathways to synthesize essential molecules, such as amino acids, nucleotides, and lipids. These pathways are crucial for bacterial growth and replication, as they provide the building blocks necessary for cell division and other cellular processes. By studying these metabolic pathways, healthcare professionals can gain a deeper understanding of how bacteria maintain their cellular functions and how they can be targeted with antimicrobial agents.

Overall, metabolic pathways play a critical role in the survival and growth of bacteria. By studying these pathways, healthcare professionals can better understand how bacteria obtain energy, synthesize essential molecules, and adapt to different environmental conditions. This knowledge is essential for developing effective strategies to combat bacterial infections and improve patient outcomes.

Energy Production

Energy production is a crucial process in bacteria that allows these microorganisms to carry out various metabolic activities necessary for their survival and growth. Bacteria utilize different mechanisms to generate energy, with the most common being through the process of cellular respiration. During cellular respiration, bacteria break down organic molecules such as glucose to produce ATP, the cell's primary energy currency.

One of the main pathways involved in energy production in bacteria is the electron transport chain. This process occurs in the bacterial cell membrane and involves the transfer of electrons from electron donors to electron acceptors, generating a proton gradient that drives ATP synthesis. The electron transport chain is essential for the production of ATP in bacteria and is a key component of their energy metabolism.

In addition to the electron transport chain, some bacteria can also generate energy through other mechanisms such as fermentation. Fermentation is a metabolic process that allows bacteria to generate ATP in the absence of oxygen by using organic molecules as electron donors and acceptors. While fermentation is less efficient than cellular respiration in terms of ATP production, it allows bacteria to survive in anaerobic environments.

Understanding the mechanisms of energy production in bacteria is essential for healthcare professionals, as it plays a crucial role in bacterial growth and survival. By targeting these energy production pathways, researchers can develop new strategies to combat bacterial infections and develop novel antimicrobial agents. Additionally, studying bacterial energy metabolism can provide valuable insights into the evolution and diversity of bacterial species.

In conclusion, energy production is a fundamental process in bacteria that drives their growth, survival, and pathogenicity. By unraveling the complexities of bacterial energy metabolism, healthcare professionals can better understand how bacteria function and develop new approaches to combat bacterial infections. This knowledge is essential for advancing our understanding of bacteriology and developing more effective treatments for bacterial diseases.

Chapter 5
Bacterial Pathogenesis

Host-Pathogen Interactions

In the field of bacteriology, one of the key areas of study is host-pathogen interactions. This subchapter delves into the complex relationship between hosts, such as humans, and pathogens, such as bacteria. Understanding these interactions is crucial for developing effective strategies for preventing and treating bacterial infections.

Host-pathogen interactions begin when a bacterium comes into contact with a host. The interaction is influenced by a variety of factors, including the virulence of the bacterium, the immune response of the host, and the environment in which the interaction takes place. Bacteria have evolved a range of mechanisms to invade and survive within host cells, while hosts have developed defense mechanisms to detect and eliminate invading pathogens.

One of the key concepts in host-pathogen interactions is the concept of pathogenicity. Pathogenicity refers to the ability of a bacterium to cause disease in a host. Factors that contribute to pathogenicity include the ability of the bacterium to adhere to host cells, invade host tissues, and evade the host's immune response. Understanding the mechanisms by which bacteria become pathogenic is essential for developing effective treatments for bacterial infections.

Another important aspect of host-pathogen interactions is the role of the host's immune response. The immune system plays a crucial role in defending the host against invading pathogens. When a bacterium enters the body, the immune system mounts a response to neutralize and eliminate the pathogen. However, bacteria have evolved strategies to evade the host's immune response, allowing them to establish infections and cause disease.

Overall, studying host-pathogen interactions is essential for understanding the mechanisms by which bacteria cause disease and for developing effective strategies for preventing and treating bacterial infections. By unraveling the complex relationship between hosts and pathogens, researchers can gain valuable insights that can be used to improve healthcare outcomes and combat the threat of antibiotic resistance.

Virulence Factors

Virulence factors are key components of bacteria that contribute to their ability to cause disease in humans. These factors allow bacteria to evade the body's immune system, adhere to host cells, and produce toxins that damage tissues. Understanding virulence factors is crucial for healthcare professionals in order to effectively diagnose and treat bacterial infections.

One important virulence factor is the ability of bacteria to produce toxins. Toxins are substances that can cause harm to the host by damaging cells or interfering with normal cellular processes. Some bacteria produce toxins that are released into the surrounding environment, while others deliver toxins directly into host cells. These toxins can cause a range of symptoms, from mild gastrointestinal distress to life-threatening organ failure.

Another important virulence factor is the ability of bacteria to adhere to host cells. By attaching to host tissues, bacteria are able to establish a foothold in the body and avoid being swept away by the immune system. Adherence factors can include proteins, carbohydrates, or other molecules that bind specifically to receptors on host cells. Once attached, bacteria can multiply and spread, leading to the development of infection.

Bacteria also possess mechanisms for evading the host immune system. This can include the ability to resist phagocytosis, a process by which immune cells engulf and destroy bacteria. Bacteria may also produce proteins that neutralize antibodies or interfere with signaling pathways that activate immune responses. By evading the immune system, bacteria are able to persist in the body and cause ongoing infection.

In addition to these factors, bacteria may also possess mechanisms for invading host tissues and spreading throughout the body. This can involve the production of enzymes that break down the extracellular matrix, allowing bacteria to move through tissues and access the bloodstream. Once in the bloodstream, bacteria can spread to other organs and tissues, causing systemic infection. By understanding the various virulence factors employed by bacteria, healthcare professionals can develop more effective strategies for preventing and treating bacterial infections.

Mechanisms of Pathogenicity

In order to understand how bacteria cause disease, it is essential to delve into the mechanisms of pathogenicity. Pathogenic bacteria possess a variety of tools and strategies that allow them to invade and colonize host tissues, evade the immune system, and cause symptoms of disease. By studying these mechanisms, healthcare professionals can gain valuable insights into how to diagnose, treat, and prevent bacterial infections.

One key mechanism of pathogenicity is the ability of bacteria to adhere to host cells. Adhesion is the first step in the process of infection, allowing bacteria to establish a foothold in the host and avoid being swept away by the body's natural defenses. Bacteria use specialized structures such as fimbriae, pili, and adhesins to bind to host cell receptors, facilitating colonization and the formation of biofilms.

Once bacteria have successfully adhered to host cells, they must then invade and replicate within host tissues. This process often involves the secretion of toxins and enzymes that disrupt host cell membranes and allow the bacteria to enter and survive within the cell. Some bacteria, such as Salmonella and Shigella, use a mechanism known as bacterial invasion to penetrate host cells and cause damage from within.

In addition to invading host tissues, pathogenic bacteria must also evade the immune system in order to establish a successful infection. Bacteria have developed a variety of strategies to evade detection and destruction by the immune system, including the production of toxins that suppress the immune response, the ability to change their surface antigens to avoid recognition by antibodies, and the formation of biofilms that shield them from immune cells.

Finally, the symptoms of bacterial infections are often the result of the host's immune response to the presence of bacteria. Inflammatory responses triggered by bacterial infection can lead to fever, swelling, pain, and tissue damage. Understanding the mechanisms by which bacteria cause disease can help healthcare professionals develop targeted therapies to combat bacterial infections and minimize the impact on the host. By studying the intricate interplay between pathogenic bacteria and their hosts, healthcare professionals can gain valuable insights into the treatment and prevention of bacterial infections.

Immune Responses to Bacterial Infections

In response to bacterial infections, the body's immune system mounts a complex and coordinated defense mechanism to combat the invading pathogens. This immune response is essential in preventing the spread of infection and promoting the body's recovery. Understanding the intricacies of immune responses to bacterial infections is crucial for healthcare professionals in diagnosing and treating patients effectively.

The immune response to bacterial infections is initiated by the recognition of the invading pathogens by the innate immune system. This recognition triggers a cascade of events that lead to the activation of immune cells, such as macrophages and dendritic cells, which engulf and destroy the bacteria. Additionally, the innate immune system releases signaling molecules called cytokines that recruit other immune cells to the site of infection to aid in the defense against the bacteria.

Following the activation of the innate immune response, the adaptive immune system is also mobilized to provide a more specific and targeted defense against the bacterial infection. This involves the activation of T cells and B cells, which work together to produce antibodies that specifically target and neutralize the bacteria. The adaptive immune response also generates memory cells that provide long-lasting immunity against future infections by the same bacteria.

In some cases, the immune response to bacterial infections can become dysregulated, leading to an overactive immune response that causes tissue damage and inflammation. This phenomenon, known as sepsis, can be life-threatening if not promptly treated. Healthcare professionals must be vigilant in monitoring patients with bacterial infections for signs of sepsis and provide appropriate interventions to prevent complications.

Overall, a thorough understanding of immune responses to bacterial infections is essential for healthcare professionals in effectively managing and treating patients with these types of infections. By recognizing the complex interplay between the innate and adaptive immune systems, healthcare providers can tailor their treatment strategies to optimize patient outcomes and promote recovery from bacterial infections.

Chapter 6
Principles of Bacterial Identification and Classification

Morphological and Biochemical Characteristics

In the subchapter "Morphological and Biochemical Characteristics" of "Essential Concepts in Bacteriology: A Primer for Healthcare Professionals," we delve into the fundamental characteristics that define bacterial species. Morphological characteristics refer to the physical appearance of bacteria, including size, shape, and arrangement. Bacteria can be classified based on these features, with common shapes including cocci (spherical), bacilli (rod-shaped), and spirilla (spiral-shaped). These morphological characteristics are essential for identifying and classifying bacteria in the laboratory.

Biochemical characteristics refer to the metabolic processes and chemical reactions that bacteria undergo. Bacteria have diverse metabolic capabilities, allowing them to utilize a wide range of nutrients for energy and growth. Some bacteria are aerobic, requiring oxygen for respiration, while others are anaerobic and can grow in the absence of oxygen. Additionally, bacteria can produce a variety of enzymes that play crucial roles in their metabolism, such as proteases, lipases, and carbohydrases.

Understanding the morphological and biochemical characteristics of bacteria is essential for healthcare professionals in diagnosing and treating bacterial infections. By identifying the shape and metabolic capabilities of a bacterial species, healthcare providers can determine the appropriate antibiotic therapy to target the infection. For example, knowing whether a bacteria is Gram-positive or Gram-negative can help guide antibiotic selection, as these two groups of bacteria have different cell wall structures that affect how they respond to antibiotics.

In addition to aiding in diagnosis and treatment, knowledge of bacterial morphology and biochemistry is crucial for preventing the spread of infections in healthcare settings. By understanding how bacteria grow and replicate, healthcare professionals can implement appropriate infection control measures to limit the transmission of pathogenic bacteria. This includes proper hand hygiene, disinfection of surfaces, and appropriate use of personal protective equipment to prevent the spread of bacteria between patients.

In conclusion, the morphological and biochemical characteristics of bacteria are essential concepts for healthcare professionals working in the field of bacteriology. By understanding these fundamental characteristics, healthcare providers can improve their ability to diagnose, treat, and prevent bacterial infections. This knowledge is vital for ensuring the safety and well-being of patients in healthcare settings and is a cornerstone of effective bacteriology practice.

Molecular Techniques

Molecular techniques have revolutionized the field of bacteriology, providing researchers and healthcare professionals with powerful tools to study and understand bacterial pathogens. These techniques allow for the identification, characterization, and manipulation of bacterial genes and proteins at the molecular level, providing valuable insights into the biology and pathogenicity of bacteria.

One of the most commonly used molecular techniques in bacteriology is polymerase chain reaction (PCR). PCR allows researchers to amplify specific DNA sequences from bacterial samples, enabling the rapid and sensitive detection of bacterial pathogens. This technique is widely used in diagnostic laboratories to identify bacterial infections and determine antibiotic resistance profiles.

Another important molecular technique in bacteriology is DNA sequencing. DNA sequencing allows for the determination of the complete genetic code of a bacterial organism, providing detailed information about its genome, including genes, regulatory sequences, and genetic variation. This information is crucial for understanding bacterial evolution, virulence factors, and mechanisms of antibiotic resistance.

In addition to PCR and DNA sequencing, other molecular techniques such as gene expression analysis, protein purification, and recombinant DNA technology are also commonly used in bacteriology. These techniques allow researchers to study the function of bacterial genes and proteins, identify potential drug targets, and develop new vaccines and therapeutics against bacterial pathogens.

Overall, molecular techniques have greatly expanded our understanding of bacterial biology and pathogenesis, leading to new insights into the mechanisms of bacterial infections and the development of novel strategies for prevention and treatment. As technology continues to advance, these techniques will play an increasingly important role in the field of bacteriology, helping to address the challenges posed by emerging infectious diseases and antibiotic resistance.

Antimicrobial Susceptibility Testing

Antimicrobial susceptibility testing is a crucial component of bacteriology, allowing healthcare professionals to determine the most effective treatment for bacterial infections. This testing involves exposing bacterial isolates to various antibiotics to assess their susceptibility or resistance to these drugs. By understanding the antimicrobial susceptibility patterns of bacterial strains, healthcare providers can make informed decisions about the best course of treatment for their patients.

There are several methods used for antimicrobial susceptibility testing, including disk diffusion, broth microdilution, and automated systems. Each method has its own advantages and limitations, and the choice of method depends on factors such as the type of bacteria being tested, available resources, and the desired turnaround time. Disk diffusion is a commonly used method that involves placing antibiotic disks on an agar plate inoculated with the test organism and measuring the zone of inhibition around each disk to determine susceptibility.

Interpreting the results of antimicrobial susceptibility testing requires an understanding of the Clinical and Laboratory Standards Institute (CLSI) guidelines, which provide breakpoints for determining whether a bacterial strain is susceptible, intermediate, or resistant to a particular antibiotic. These guidelines are updated regularly to reflect changes in bacterial resistance patterns and ensure that healthcare providers have the most up-to-date information for guiding treatment decisions. It is essential for healthcare professionals to stay informed about these guidelines to ensure that patients receive the most effective antibiotic therapy.

Antimicrobial susceptibility testing plays a critical role in guiding antibiotic therapy and preventing the spread of antibiotic-resistant bacteria. By conducting this testing, healthcare professionals can identify the most effective antibiotics for treating bacterial infections, reducing the risk of treatment failure and the development of resistance. Additionally, antimicrobial susceptibility testing allows for the monitoring of resistance trends in bacterial populations, informing public health efforts to combat antibiotic resistance on a larger scale.

In conclusion, antimicrobial susceptibility testing is a vital tool in the field of bacteriology, enabling healthcare professionals to make evidence-based decisions about antibiotic therapy. By understanding the methods, interpretation guidelines, and clinical implications of this testing, healthcare providers can optimize patient care and contribute to the global effort to combat antibiotic resistance. Staying informed about

A Primer for Healthcare Professionals

the latest developments in antimicrobial susceptibility testing is essential for healthcare professionals working in the field of bacteriology.

Chapter 7
Antibiotics and Antibiotic Resistance

Mechanisms of Action

In the field of bacteriology, understanding the mechanisms of action of bacteria is crucial for healthcare professionals to effectively combat bacterial infections. This subchapter will delve into the various ways in which bacteria exert their effects on the human body, providing readers with a comprehensive overview of these key processes.

One of the primary mechanisms of action employed by bacteria is the production of toxins. Toxins are substances released by bacteria that can cause harm to the host by damaging cells, tissues, and organs. Some bacteria produce exotoxins, which are released into the surrounding environment, while others produce endotoxins, which are released when the bacteria are killed. Understanding how these toxins work is essential for developing treatments to combat bacterial infections.

Another important mechanism of action used by bacteria is their ability to evade the host's immune system. Bacteria have developed a variety of strategies to avoid detection and destruction by the immune system, allowing them to establish infections and replicate within the body. By understanding these evasion mechanisms, healthcare professionals can develop targeted therapies to enhance the immune response and improve patient outcomes.

Bacteria also have the ability to form biofilms, which are complex communities of bacteria encased in a protective matrix. Biofilms allow bacteria to adhere to surfaces, evade immune responses, and resist antibiotics, making them challenging to treat. Understanding how bacteria form and maintain biofilms is essential for developing strategies to disrupt these structures and improve treatment outcomes for patients with biofilm-associated infections.

In addition to toxins, immune evasion, and biofilm formation, bacteria also employ a variety of other mechanisms of action to colonize and cause disease in the human body. By studying these mechanisms, healthcare professionals can gain valuable insights into how bacteria interact with their hosts and develop new approaches to prevent and treat bacterial infections. This subchapter will provide readers with a solid foundation in the mechanisms of action of bacteria, enabling them to better understand and address the challenges posed by these diverse and adaptable microorganisms.

Antibiotic Classes

Antibiotics are essential tools in the fight against bacterial infections. They work by targeting specific components of bacterial cells, disrupting their ability to grow and reproduce. There are several different classes of antibiotics, each with its own unique mechanism of action and spectrum of activity. Understanding the various antibiotic classes is crucial for healthcare professionals in order to effectively treat bacterial infections and prevent the development of antibiotic resistance.

One of the most commonly used classes of antibiotics is the beta-lactams, which includes penicillins, cephalosporins, and carbapenems. These antibiotics work by inhibiting the synthesis of the bacterial cell wall, leading to cell death. Beta-lactams are effective against a wide range of gram-positive and gram-negative bacteria, making them a versatile choice for many infections.

Another important class of antibiotics is the macrolides, which includes erythromycin, azithromycin, and clarithromycin. Macrolides work by interfering with bacterial protein synthesis, preventing the bacteria from multiplying. They are often used to treat respiratory tract infections, skin infections, and sexually transmitted diseases.

Fluoroquinolones are a class of antibiotics that target bacterial DNA replication, preventing the bacteria from reproducing. Ciprofloxacin and levofloxacin are examples of fluoroquinolones that are commonly used to treat urinary tract infections, respiratory infections, and skin infections.

Tetracyclines are a class of antibiotics that inhibit bacterial protein synthesis, similar to macrolides. They are effective against a wide range of bacteria, including some that are resistant to other classes of antibiotics. Tetracyclines are commonly used to treat acne, respiratory infections, and sexually transmitted diseases.

In conclusion, understanding the various classes of antibiotics is essential for healthcare professionals in order to effectively treat bacterial infections and prevent the development of antibiotic resistance. By knowing how each class of antibiotics works and their spectrum of activity, healthcare professionals can make informed decisions when selecting the appropriate antibiotic for a specific infection. It is important to use antibiotics judiciously and follow proper prescribing practices to ensure their continued effectiveness in combating bacterial infections.

Factors Contributing to Antibiotic Resistance

Antibiotic resistance is a growing concern in the field of bacteriology, posing a significant threat to public health worldwide. Understanding the factors contributing to antibiotic resistance is crucial in developing strategies to combat this issue effectively. Several key factors play a role in the development and spread of antibiotic resistance among bacterial populations.

One of the primary factors contributing to antibiotic resistance is the overuse and misuse of antibiotics. The widespread use of antibiotics in healthcare settings, agriculture, and veterinary medicine has led to the emergence of resistant strains of bacteria. Overprescription of antibiotics, failure to complete prescribed courses, and inappropriate use of broad-spectrum antibiotics all contribute to the development of resistance.

Another important factor is the misuse of antibiotics in animal agriculture. Antibiotics are commonly used in livestock farming to promote growth and prevent disease. The widespread use of antibiotics in this setting has led to the development of resistant bacteria that can be transmitted to humans through the food chain. This practice has been a major contributor to the rise of antibiotic-resistant infections in humans.

The spread of antibiotic resistance is also facilitated by poor infection control practices in healthcare settings. Inadequate hand hygiene, improper sterilization of medical equipment, and overcrowding in hospitals all contribute to the transmission of resistant bacteria among patients. Healthcare-associated infections caused by antibiotic-resistant bacteria are a significant challenge for healthcare professionals, requiring strict adherence to infection control protocols to prevent their spread.

In addition to these factors, the lack of new antibiotics in development is a major concern in the fight against antibiotic resistance. The pipeline for new antibiotic drugs has been dwindling in recent years, with few new drugs being brought to market. This lack of innovation leaves healthcare professionals with limited treatment options for infections caused by resistant bacteria, highlighting the urgent need for new strategies to combat antibiotic resistance.

Overall, a combination of factors contributes to the development and spread of antibiotic resistance among bacterial populations. Addressing these factors through effective antibiotic stewardship programs, improved infection control practices, and increased research and development of new antibiotics is essential in combating this growing public health threat.

Strategies to Combat Antibiotic Resistance

Antibiotic resistance is a growing concern in healthcare today, as bacteria evolve and develop resistance to the drugs that are meant to fight them. In order to combat this issue, healthcare professionals must employ various strategies to preserve the effectiveness of antibiotics. One key strategy is the judicious use of antibiotics. This means only prescribing antibiotics when necessary and ensuring that they are used correctly and for the appropriate duration. Overuse and misuse of antibiotics can contribute to the development of resistance, so it is crucial for healthcare providers to be mindful of their prescribing practices.

Another important strategy to combat antibiotic resistance is the practice of antibiotic stewardship. This involves implementing policies and procedures to optimize antibiotic use in order to improve patient outcomes and reduce the spread of resistance. By promoting the appropriate use of antibiotics and monitoring their effectiveness, healthcare facilities can help to slow the development of resistance and preserve the efficacy of these life-saving drugs.

In addition to judicious use and stewardship, healthcare professionals can also combat antibiotic resistance by promoting infection prevention and control measures. By implementing practices such as hand hygiene, proper sanitation, and isolation precautions, healthcare facilities can reduce the spread of resistant bacteria and prevent infections from occurring in the first place. These measures are essential in protecting both patients and healthcare workers from the dangers of antibiotic-resistant infections.

Furthermore, education plays a crucial role in combating antibiotic resistance. Healthcare professionals must stay informed about the latest developments in antimicrobial resistance and antibiotic stewardship in order to effectively address this issue in their practice. By educating themselves and their patients about the proper use of antibiotics and the importance of preventing resistance, healthcare providers can help to curb the spread of resistant bacteria and preserve the effectiveness of these vital drugs.

Overall, combating antibiotic resistance requires a multi-faceted approach that involves judicious use of antibiotics, antibiotic stewardship, infection prevention and control measures, and ongoing education. By implementing these strategies in healthcare settings, healthcare professionals can help to slow the development of antibiotic resistance and ensure that antibiotics remain effective for future generations.

Chapter 8
Bacterial Diseases

Gram-Positive Bacterial Infections

Gram-positive bacterial infections are a significant concern in healthcare settings, as these organisms can cause a wide range of illnesses, from mild skin infections to life-threatening systemic diseases. Gram-positive bacteria are characterized by their thick peptidoglycan cell wall, which retains the crystal violet dye during the Gram staining process. This cell wall structure gives these bacteria their unique appearance under the microscope and also plays a role in their pathogenicity.

One of the most well-known gram-positive bacteria is Staphylococcus aureus, which is a common cause of skin and soft tissue infections. This bacterium can also cause more serious infections, such as pneumonia, endocarditis, and sepsis. Another important gram-positive pathogen is Streptococcus pyogenes, which is responsible for a wide range of infections, including strep throat, cellulitis, and necrotizing fasciitis. These bacteria are typically treated with antibiotics such as penicillin or vancomycin.

In recent years, there has been a growing concern about the rise of antibiotic-resistant gram-positive bacteria, such as methicillin-resistant Staphylococcus aureus (MRSA) and vancomycin-resistant Enterococcus (VRE). These bacteria pose a significant challenge to healthcare providers, as they are often difficult to treat with standard antibiotics. Strategies to combat antibiotic resistance include appropriate antibiotic stewardship, infection control measures, and the development of new antibiotics.

Healthcare professionals must be vigilant in identifying and treating gram-positive bacterial infections to prevent the spread of these pathogens and reduce the risk of complications. Proper hand hygiene, appropriate use of antibiotics, and infection control protocols are essential in controlling the spread of gram-positive bacteria in healthcare settings. By understanding the mechanisms of gram-positive bacterial infections and implementing appropriate prevention and treatment strategies, healthcare professionals can help to reduce the burden of these infections on patients and healthcare systems.

In conclusion, gram-positive bacterial infections are a significant concern in healthcare settings, and healthcare professionals must be knowledgeable about the pathogens involved, their mechanisms of infection, and appropriate treatment strategies. By staying informed about the latest developments in bacteriology and taking proactive measures to prevent the spread of these infections, healthcare providers can help to protect their patients and communities from the harmful effects of gram-positive bacteria.

Gram-Negative Bacterial Infections

Gram-negative bacterial infections are a significant concern in healthcare settings due to their resistance to many antibiotics and ability to cause serious illnesses. These types of bacteria have a distinct cell wall structure that makes them more difficult to treat than gram-positive bacteria. Understanding the characteristics and treatment options for gram-negative bacterial infections is essential for healthcare professionals to effectively manage and prevent the spread of these pathogens.

One key feature of gram-negative bacteria is their outer membrane, which contains lipopolysaccharides that can trigger a strong immune response in the host. This can lead to symptoms such as fever, chills, and inflammation. Common gram-negative bacterial infections include E. coli, Pseudomonas aeruginosa, and Klebsiella pneumoniae, which can cause a range of illnesses from urinary tract infections to pneumonia.

Treatment of gram-negative bacterial infections can be challenging due to the limited number of effective antibiotics available. Healthcare professionals must carefully choose the appropriate antibiotic based on the specific bacteria causing the infection and its susceptibility to different drugs. In some cases, combination therapy may be necessary to improve treatment outcomes and prevent the development of antibiotic resistance.

Preventing the spread of gram-negative bacterial infections is crucial in healthcare settings to protect patients and healthcare workers. This includes implementing strict infection control measures such as hand hygiene, proper use of personal protective equipment, and disinfection of medical equipment and surfaces. Additionally, surveillance of antibiotic resistance patterns can help identify emerging threats and guide treatment strategies.

In conclusion, gram-negative bacterial infections present unique challenges for healthcare professionals due to their resistance to many antibiotics and ability to cause serious illnesses. Understanding the characteristics of these pathogens, choosing appropriate treatment options, and implementing effective infection control measures are essential for managing and preventing the spread of gram-negative bacterial infections in healthcare settings. By staying informed and proactive, healthcare professionals can help reduce the impact of these pathogens on patient outcomes and public health.

Atypical Bacterial Infections

Atypical bacterial infections are caused by bacteria that do not fit into the typical categories of pathogenic organisms. These infections can be difficult to diagnose and treat, as they may present with unusual symptoms or be resistant to standard treatments. In this subchapter, we will explore some of the most common atypical bacterial infections encountered in healthcare settings.

One of the most well-known atypical bacterial infections is Mycoplasma pneumoniae, which causes a form of pneumonia known as "walking pneumonia." This bacterium lacks a cell wall, making it resistant to many antibiotics that target cell wall synthesis. Patients with Mycoplasma pneumoniae infections often present with mild respiratory symptoms that may be mistaken for a cold or flu, making diagnosis challenging.

Another atypical bacterial infection is caused by Chlamydia trachomatis, which can lead to a range of conditions including genital infections, pneumonia, and conjunctivitis. This bacterium is an obligate intracellular parasite, meaning it can only survive and replicate inside host cells. Chlamydia infections are typically treated with antibiotics, but strains resistant to common therapies are emerging, posing a challenge to healthcare providers.

Legionella pneumophila is another atypical bacterial pathogen that can cause severe pneumonia known as Legionnaires' disease. This bacterium thrives in water systems such as cooling towers and hot tubs, making outbreaks more common in certain environments. Legionella infections can be difficult to diagnose, as they often present with symptoms similar to other types of pneumonia. Prompt treatment with antibiotics is essential to prevent severe complications.

Finally, Rickettsia rickettsii is a bacterium that causes Rocky Mountain spotted fever, a potentially life-threatening infection transmitted by ticks. This atypical bacterial infection presents with symptoms such as fever, rash, and headache, and can progress rapidly if not treated promptly. Healthcare providers must be vigilant in recognizing the signs of Rocky Mountain spotted fever and initiating appropriate antibiotic therapy to prevent serious complications.

In conclusion, atypical bacterial infections present unique challenges for healthcare professionals due to their unusual characteristics and resistance to standard treatments. Understanding the key features of these pathogens and staying informed about emerging resistant strains is essential for effective diagnosis and management. By remaining vigilant and proactive in identifying atypical bacterial infections, healthcare providers can ensure the best possible outcomes for their patients.

Emerging Infectious Diseases

Emerging infectious diseases are a growing concern in the field of bacteriology. These diseases are caused by newly identified pathogens or by known pathogens that have evolved to become more virulent or resistant to current treatments. The rapid spread of emerging infectious diseases poses a significant threat to public health and highlights the importance of ongoing research and surveillance efforts.

One example of an emerging infectious disease is the Zika virus, which gained global attention in 2015. The virus, transmitted by Aedes mosquitoes, caused outbreaks of microcephaly and Guillain-Barré syndrome in several countries. The sudden emergence of Zika virus highlighted the need for improved mosquito control measures and increased research into the mechanisms of viral transmission and pathogenesis.

Another example of an emerging infectious disease is multidrug-resistant tuberculosis (MDR-TB). This form of tuberculosis is resistant to at least two of the most powerful first-line anti-TB drugs, isoniazid and rifampicin. MDR-TB is a significant challenge for healthcare professionals as it requires longer and more complex treatment regimens, often with less effective and more toxic drugs. The emergence of MDR-TB underscores the importance of antimicrobial stewardship and infection control practices in healthcare settings.

In addition to viral and bacterial pathogens, emerging infectious diseases can also be caused by fungi, parasites, and other microorganisms. For example, the emergence of Candida auris, a multidrug-resistant fungus, has led to outbreaks of invasive infections in healthcare settings worldwide. The ability of C. auris to persist in the environment and on surfaces for weeks highlights the need for strict infection control measures to prevent its spread.

In conclusion, emerging infectious diseases represent a significant challenge for healthcare professionals and public health authorities. Ongoing research, surveillance, and collaboration are essential to monitor and control the spread of these diseases. By staying informed and implementing effective prevention and control measures, healthcare professionals can help mitigate the impact of emerging infectious diseases on global health.

Chapter 9
Bacterial Control and Prevention

Sterilization and Disinfection

Sterilization and disinfection are two essential processes in the field of bacteriology that are utilized to prevent the spread of harmful microorganisms and maintain a safe and hygienic environment in healthcare settings. Sterilization refers to the complete destruction or removal of all forms of microbial life, including bacteria, viruses, fungi, and spores. This process is crucial in healthcare facilities to ensure that medical instruments, equipment, and surfaces are free from any potential pathogens that could cause infections.

There are several methods of sterilization that are commonly used in healthcare settings, including autoclaving, dry heat sterilization, ethylene oxide gas sterilization, and gamma irradiation. Autoclaving, which involves the use of steam under pressure, is one of the most effective and widely used methods for sterilizing medical instruments and equipment. Dry heat sterilization, on the other hand, utilizes high temperatures to kill microorganisms and is often used for items that are sensitive to moisture.

Disinfection, on the other hand, refers to the process of reducing the number of microorganisms on surfaces to a level that is considered safe for human health. Disinfectants are chemical agents that are used to kill or inactivate microorganisms on surfaces and are commonly used in healthcare settings to clean and disinfect patient rooms, operating rooms, and other high-touch surfaces. It is important to note that disinfection is not as effective as sterilization at eliminating all forms of microbial life, but it is still a crucial step in preventing the spread of infections.

When choosing a disinfectant for use in healthcare settings, it is important to consider factors such as the type of microorganisms present, the level of disinfection required, and the compatibility of the disinfectant with the surfaces being cleaned. Different disinfectants have varying levels of efficacy against different types of microorganisms, so it is important to select the appropriate disinfectant for the specific situation. Additionally, it is important to follow the manufacturer's instructions for use, including the contact time required for the disinfectant to be effective.

In conclusion, sterilization and disinfection are essential processes in the field of bacteriology that are crucial for maintaining a safe and hygienic environment in healthcare settings. By understanding the different methods of sterilization and disinfection and selecting the appropriate techniques and disinfectants for each situation, healthcare professionals can help prevent the spread of infections and ensure the safety of patients and staff.

Hand Hygiene

Hand hygiene is a critical aspect of infection control in healthcare settings. Proper hand hygiene practices are essential for preventing the spread of harmful bacteria and viruses that can cause infections. Healthcare professionals must adhere to strict hand hygiene protocols to protect both themselves and their patients from the transmission of pathogens.

The most effective method of hand hygiene is washing hands with soap and water. The friction created by rubbing hands together while washing helps to physically remove dirt, bacteria, and viruses from the skin. Healthcare professionals should wash their hands before and after patient contact, after touching contaminated surfaces, and before and after performing invasive procedures. It is important to wash hands for at least 20 seconds to ensure thorough cleaning.

In situations where soap and water are not readily available, healthcare professionals can use alcohol-based hand sanitizers. These sanitizers are effective at killing a broad spectrum of bacteria and viruses, including the ones that cause healthcare-associated infections. However, it is important to note that hand sanitizers are not effective against all types of pathogens, such as norovirus and Clostridium difficile. In these cases, washing hands with soap and water is necessary.

In addition to hand washing and hand sanitizing, healthcare professionals should also practice proper hand drying techniques. Wet hands can transfer bacteria more easily than dry hands, so it is important to thoroughly dry hands with a clean towel or air dryer after washing. Healthcare facilities should provide adequate supplies of soap, water, hand sanitizer, and hand drying materials to ensure that staff can maintain proper hand hygiene at all times.

Overall, hand hygiene is a simple yet effective way to prevent the spread of infections in healthcare settings. By following proper hand hygiene protocols, healthcare professionals can protect themselves, their patients, and their colleagues from harmful bacteria and viruses. It is essential for all healthcare professionals to make hand hygiene a priority in their daily practice to ensure the safety and well-being of everyone in their care.

Vaccination

Vaccination is a crucial aspect of public health and plays a significant role in preventing the spread of infectious diseases caused by bacteria. By stimulating the immune system to produce antibodies against specific pathogens, vaccines can effectively protect individuals from getting sick and help to create herd immunity within a population. This subchapter will explore the history of vaccination, the different types of vaccines available, and the importance of vaccination in the context of bacteriology.

The practice of vaccination dates back to the late 18th century when Edward Jenner developed the smallpox vaccine. This breakthrough paved the way for the development of vaccines against other bacterial diseases such as tetanus, diphtheria, and pertussis. Today, vaccines are considered one of the most successful and cost-effective public health interventions, saving millions of lives each year.

There are several types of vaccines used to protect against bacterial diseases, including live attenuated vaccines, inactivated vaccines, subunit vaccines, and conjugate vaccines. Each type of vaccine works in a slightly different way to stimulate the immune system and provide protection against specific bacteria. For example, live attenuated vaccines contain weakened forms of the pathogen, while inactivated vaccines contain killed versions of the bacteria.

Vaccination is essential in the field of bacteriology because it helps to prevent the spread of bacterial infections and reduce the burden of disease on healthcare systems. By vaccinating individuals against common bacterial pathogens, healthcare professionals can effectively control outbreaks and protect vulnerable populations, such as infants, the elderly, and individuals with weakened immune systems. Additionally, vaccination can help to prevent the emergence of antibiotic-resistant bacteria by reducing the need for antibiotics to treat infections.

In conclusion, vaccination is a vital tool in the fight against bacterial diseases and plays a crucial role in maintaining public health. By understanding the history of vaccination, the different types of vaccines available, and the importance of vaccination in bacteriology, healthcare professionals can work together to promote vaccination as a key strategy for preventing the spread of infectious diseases caused by bacteria.

Infection Control Measures

Infection control measures are essential in healthcare settings to prevent the spread of harmful bacteria and other pathogens. By implementing proper infection control practices, healthcare professionals can protect both themselves and their patients from the transmission of infectious diseases. In this section, we will discuss some of the key infection control measures that should be followed in all healthcare settings.

One of the most important infection control measures is hand hygiene. Healthcare professionals should always wash their hands with soap and water or use hand sanitizer before and after patient contact. This simple practice can help prevent the spread of bacteria and viruses that can cause infections. Additionally, healthcare facilities should provide easy access to hand hygiene products and promote hand hygiene compliance among staff.

Another important infection control measure is the proper use of personal protective equipment (PPE). Healthcare workers should wear gloves, gowns, masks, and eye protection when caring for patients with infectious diseases. By using PPE correctly, healthcare professionals can protect themselves and their patients from exposure to harmful pathogens. It is essential for healthcare facilities to provide appropriate PPE and ensure that staff are trained on how to use it effectively.

Environmental cleaning is also a crucial infection control measure. Healthcare facilities should have protocols in place for the cleaning and disinfection of patient care areas, equipment, and high-touch surfaces. Regular cleaning can help prevent the spread of bacteria and other pathogens that can cause infections. Healthcare facilities should also have procedures for handling and disposing of infectious waste properly to minimize the risk of contamination.

Infection control measures should also include protocols for the safe handling and disposal of sharps, such as needles and syringes. Healthcare professionals should always use safety devices when handling sharps to prevent needlestick injuries and the spread of bloodborne pathogens. Proper sharps disposal is essential to protect healthcare workers, patients, and the community from the risks associated with needlestick injuries and exposure to infectious diseases.

Overall, infection control measures are critical in healthcare settings to prevent the spread of infectious diseases. By following proper hand hygiene practices, using personal protective equipment, maintaining a clean environment, and safely handling sharps, healthcare professionals can help protect themselves and their patients from the transmission of harmful bacteria and other pathogens. It is essential for healthcare facilities to have infection control protocols in place and ensure that all staff are trained on how to implement them effectively.

Chapter 10
Future Directions in Bacteriology

Advances in Bacterial Research

Advances in Bacterial Research have revolutionized our understanding of these microscopic organisms and their impact on human health. From the discovery of new bacterial species to the development of novel treatment strategies, researchers are constantly making groundbreaking discoveries in the field of bacteriology. These advances have paved the way for new diagnostic tools, antibiotics, and vaccines that have the potential to save countless lives and improve public health outcomes.

One of the most exciting advancements in bacterial research is the use of genomic sequencing to identify and characterize bacterial strains. By analyzing the DNA of bacteria, researchers can gain valuable insights into their genetic makeup, virulence factors, and antibiotic resistance mechanisms. This information is crucial for developing targeted treatment strategies and preventing the spread of drug-resistant bacteria in healthcare settings.

Another major breakthrough in bacterial research is the development of phage therapy as an alternative to antibiotics. Bacteriophages are viruses that specifically target and kill bacteria, offering a promising new approach to treating bacterial infections. Researchers are now exploring the potential of phage therapy to combat antibiotic-resistant bacteria and improve patient outcomes in the era of antibiotic resistance.

In addition to new treatment options, advances in bacterial research have also led to the discovery of novel bacterial species that play important roles in human health. For example, recent studies have identified beneficial bacteria in the gut microbiome that contribute to digestion, immune function, and overall well-being. Understanding the complex interactions between bacteria and the human body is essential for developing personalized therapies that promote health and prevent disease.

Overall, the rapid pace of advances in bacterial research holds great promise for improving healthcare outcomes and addressing the challenges posed by antibiotic resistance. By leveraging cutting-edge technologies and innovative approaches, researchers are uncovering new insights into the biology of bacteria and developing novel strategies to combat infectious diseases. As we continue to expand our knowledge of bacterial pathogens and their mechanisms of action, we are better equipped to tackle the threats they pose to public health and develop effective interventions to protect human health.

Potential Therapeutic Strategies

In the field of bacteriology, the development of potential therapeutic strategies is crucial in combating bacterial infections and diseases. There are several approaches that healthcare professionals can consider when treating bacterial infections, ranging from traditional antibiotics to cutting-edge therapies. Understanding these strategies is essential for providing effective treatment and preventing the spread of drug-resistant bacteria.

One of the most common therapeutic strategies for bacterial infections is the use of antibiotics. Antibiotics work by targeting specific components of bacterial cells, such as cell walls or protein synthesis machinery, to inhibit their growth and replication. However, overuse and misuse of antibiotics have led to the emergence of drug-resistant bacteria, making it increasingly challenging to treat infections. Healthcare professionals must be aware of the proper use of antibiotics and consider alternatives when necessary.

In recent years, there has been a growing interest in the development of alternative therapies for bacterial infections. One promising approach is the use of bacteriophages, which are viruses that infect and kill specific bacterial strains. Bacteriophages offer a targeted and potentially more effective treatment option compared to broad-spectrum antibiotics. Research into bacteriophage therapy is ongoing, with promising results in animal studies and clinical trials.

Another potential therapeutic strategy for bacterial infections is the use of probiotics. Probiotics are live microorganisms that can confer health benefits when consumed in adequate amounts. By promoting the growth of beneficial bacteria in the gut, probiotics can help maintain a healthy balance of microorganisms and prevent the overgrowth of harmful bacteria. Healthcare professionals can consider incorporating probiotics into treatment plans for certain bacterial infections, such as Clostridium difficile-associated diarrhea.

In addition to traditional antibiotics, bacteriophages, and probiotics, healthcare professionals can explore other innovative approaches to treating bacterial infections. These may include immunotherapy, nanotechnology, and biofilm-disrupting agents. By staying informed about the latest research and developments in the field of bacteriology, healthcare professionals can tailor their treatment strategies to provide the best possible care for patients with bacterial infections.

Global Health Implications of Bacterial Infections

Bacterial infections play a significant role in global health, affecting millions of individuals each year. The implications of these infections are far-reaching, impacting not only the individuals who are directly affected but also their communities and even entire populations. In this subchapter, we will explore the global health implications of bacterial infections and the challenges they present to healthcare professionals worldwide.

One of the primary concerns regarding bacterial infections is the rise of antibiotic resistance. As bacteria evolve and develop resistance to commonly used antibiotics, treating infections becomes increasingly difficult. This poses a serious threat to global health, as antibiotic-resistant bacteria can spread quickly and cause widespread outbreaks that are difficult to control. Healthcare professionals must stay vigilant and adapt their treatment strategies to combat this growing problem.

In addition to antibiotic resistance, bacterial infections can also lead to severe complications and long-term health consequences. For example, untreated bacterial infections can progress to more serious conditions such as sepsis, which can be life-threatening if not promptly treated. In some cases, bacterial infections can also cause chronic conditions such as arthritis or heart disease, further complicating the health outcomes for affected individuals.

The global impact of bacterial infections is especially significant in low-resource settings, where access to healthcare and appropriate treatment options may be limited. In these regions, bacterial infections can lead to high rates of morbidity and mortality, particularly among vulnerable populations such as children and the elderly. Healthcare professionals working in these areas face unique challenges in diagnosing and treating bacterial infections, requiring innovative solutions to improve health outcomes.

Overall, understanding the global health implications of bacterial infections is essential for healthcare professionals in order to effectively prevent, diagnose, and treat these infections. By staying informed about emerging trends in bacterial resistance, recognizing the potential complications of untreated infections, and addressing the unique challenges faced in low-resource settings, healthcare professionals can work towards improving global health outcomes related to bacterial infections.

Chapter 11
Conclusion

Summary of Key Concepts

In this subchapter titled "Summary of Key Concepts," we will review the essential concepts covered in this book, "Essential Concepts in Bacteriology: A Primer for Healthcare Professionals." Bacteriology is the study of bacteria, which are microscopic organisms that can be found in various environments, including soil, water, and the human body. Understanding the basic concepts of bacteriology is crucial for healthcare professionals to diagnose and treat bacterial infections effectively.

One key concept covered in this book is the structure and classification of bacteria. Bacteria can be classified based on their shape, size, and cell wall composition. Understanding the different types of bacteria can help healthcare professionals identify the specific bacteria causing an infection and choose the appropriate treatment.

Another important concept discussed in this book is the mechanisms of bacterial pathogenesis. Bacteria can cause disease by producing toxins, invading host cells, and triggering an immune response. Understanding how bacteria cause disease is essential for developing strategies to prevent and treat bacterial infections.

The book also covers the principles of antimicrobial therapy, which is the use of medications to treat bacterial infections. Healthcare professionals must understand the mechanisms of action of antibiotics, as well as factors that can lead to antibiotic resistance. Proper use of antibiotics is crucial to prevent the spread of resistant bacteria and ensure effective treatment of bacterial infections.

Lastly, this book explores the role of bacteriology in public health. Bacteria can cause outbreaks of infectious diseases, leading to serious public health concerns. Healthcare professionals must be aware of the principles of infection control, surveillance, and outbreak investigation to prevent the spread of bacterial infections and protect the health of the community.

In conclusion, "Essential Concepts in Bacteriology: A Primer for Healthcare Professionals" provides a comprehensive overview of key concepts in bacteriology that are essential for healthcare professionals. By understanding the structure and classification of bacteria, mechanisms of bacterial pathogenesis, principles of antimicrobial therapy, and role of bacteriology in public health, healthcare professionals can effectively diagnose, treat, and prevent bacterial infections.

Recommendations for Further Reading

For readers interested in delving deeper into the world of bacteriology, there are numerous resources available that can provide additional insight and information on the subject. Below are some recommendations for further reading that can help expand your knowledge and understanding of essential concepts in bacteriology.

1. "Brock Biology of Microorganisms" by Michael T. Madigan, John M. Martinko, and Kelly S. Bender: This comprehensive textbook covers a wide range of topics in microbiology, including bacteriology. It provides detailed explanations of key concepts and principles, making it a valuable resource for healthcare professionals looking to deepen their understanding of the field.

2. "Medical Microbiology" by Patrick R. Murray, Ken S. Rosenthal, and Michael A. Pfaller: This textbook offers a comprehensive overview of medical microbiology, including bacteriology. It covers the latest research and developments in the field, making it a valuable resource for healthcare professionals seeking to stay up-to-date on the latest advancements in bacteriology.

3. "Principles of Bacteriology, Virology, and Immunity" by Ronald M. Atlas: This textbook provides a detailed exploration of the fundamental principles of bacteriology, virology, and immunity. It offers a comprehensive overview of the key concepts in bacteriology, making it an essential resource for healthcare professionals looking to deepen their knowledge of the field.

4. "Clinical Microbiology Made Ridiculously Simple" by Mark Gladwin and William Trattler: This book provides a simplified yet comprehensive overview of clinical microbiology, including bacteriology. It uses a conversational tone and engaging illustrations to help readers grasp complex concepts more easily, making it a popular choice for healthcare professionals seeking a user-friendly resource on bacteriology.

5. "Mims' Medical Microbiology" by Richard Goering, Hazel Dockrell, Mark Zuckerman, and Peter L. Chiodini: This textbook offers a comprehensive overview of medical microbiology, including bacteriology. It covers a wide range of topics in the field, making it a valuable resource for healthcare professionals looking to deepen their understanding of essential concepts in bacteriology.

Importance of Bacteriology in Healthcare Practice

Bacteriology plays a crucial role in healthcare practice by helping healthcare professionals understand the causes of infectious diseases and how to effectively prevent and treat them. Understanding the behavior and characteristics of bacteria is essential for diagnosing and treating bacterial infections. By studying bacteriology, healthcare professionals can identify the specific bacteria responsible for an infection and determine the most appropriate course of treatment.

One of the key reasons why bacteriology is important in healthcare practice is because it helps in the development of antibiotics and other antimicrobial agents. Antibiotics are essential in treating bacterial infections, and without a solid understanding of bacteriology, it would be impossible to develop effective antibiotics. Bacteriology also plays a role in the development of vaccines, which are essential for preventing bacterial infections. By studying bacteriology, healthcare professionals can stay informed about the latest developments in antibiotic resistance and work towards developing new treatments to combat resistant bacteria.

Another reason why bacteriology is important in healthcare practice is because it helps in infection control and prevention. By understanding how bacteria spread and cause infections, healthcare professionals can implement effective infection control measures to prevent the spread of bacterial infections in healthcare settings. Bacteriology also plays a role in public health, as healthcare professionals can use their knowledge of bacteria to track and control outbreaks of infectious diseases in communities.

Furthermore, bacteriology is important in healthcare practice because it helps in the diagnosis of bacterial infections. By studying the characteristics of different bacteria, healthcare professionals can use laboratory tests to identify the specific bacteria causing an infection. This information is essential for determining the most appropriate treatment for the infection and preventing its spread to others. Bacteriology also plays a role in research and development, as healthcare professionals can use their knowledge of bacteria to develop new diagnostic tests and treatment strategies for bacterial infections.

In conclusion, bacteriology is an essential field of study for healthcare professionals as it provides the foundation for understanding and treating bacterial infections. By studying bacteriology, healthcare professionals can develop effective treatments for bacterial infections, prevent the spread of infections, and improve public health. Bacteriology also plays a crucial role in research and development, as healthcare professionals work towards developing new antibiotics and treatment strategies to combat bacterial infections. Overall, a solid understanding of bacteriology is essential for healthcare professionals in providing quality care to patients and preventing the spread of infectious diseases.

Dear Reader,

Thank you for choosing to read "Essential Concepts in Bacteriology: A Primer for Healthcare Professionals." I hope this book has provided you with valuable knowledge and a deeper understanding of bacteriology, essential for your professional practice.

Your feedback is incredibly important to me and to other healthcare professionals seeking reliable resources. If you found this book helpful or informative, I would greatly appreciate it if you could take a moment to leave a review on Amazon. Your honest review will help others discover the book and understand its value.

To leave a review, simply follow these steps:
1. Go to the book's page on Amazon.
2. Scroll down to the "Customer Reviews" section.
3. Click on "Write a customer review."
4. Share your thoughts and experiences with the book.

Thank you so much for your time and support. Your review means a lot to me and to the healthcare community!

Best regards,
Bhupen Thapa

www.ingramcontent.com/pod-product-compliance
Lightning Source LLC
Chambersburg PA
CBHW072019230526
45479CB00008B/301